Mastering 11+

Maths

Numerical Reasoning

Practice Book 1

ashkraft
EDUCATIONAL

This page is intentionally left blank

Mastering 11+
Maths / Numerical Reasoning
Practice Book 1

Copyright © 2014 ASHKRAFT EDUCATIONAL

ISBN: 1503089894
ISBN-13: 978-1503089891

9 781503 089891 >

DEDICATION

All the little mathematicians practicing for the
Eleven plus challenge.

"Do your work with your whole heart and
you will succeed – there's so little competition"

Elbert Hubbard

Table of Contents

EXERCISE 1: Short Numerical Reasoning	1
EXERCISE 2: Short Numerical Reasoning	6
EXERCISE 3: Short Numerical Reasoning	10
EXERCISE 4: Short Numerical Reasoning	14
EXERCISE 5: Short Numerical Reasoning	18
EXERCISE 6: Short Numerical Reasoning	22
EXERCISE 7: Short Numerical Reasoning	26
EXERCISE 8: Short Numerical Reasoning	30
EXERCISE 9: Short Numerical Reasoning	34
EXERCISE 10: Short Numerical Reasoning	38
EXERCISE 11: Multi Part Numerical Reasoning	43
QUESTION - ONE	43
QUESTION - TWO	44
QUESTION – THREE	46
QUESTION – FOUR	47
Answer Sheet – Exercise 11	48
EXERCISE 12: Multi Part Numerical Reasoning	52
QUESTION - ONE	52
QUESTION - TWO	53
QUESTION - THREE	54
QUESTION – FOUR	56
Answer Sheet – Exercise 12	57
EXERCISE 13: Multi Part Numerical Reasoning	61

QUESTION – ONE	61
QUESTION – TWO	62
QUESTION – THREE	63
QUESTION – FOUR	65
Answer Sheet – Exercise 13	66
EXERCISE 14: Multi Part Numerical Reasoning	70
QUESTION – ONE	70
QUESTION - TWO	72
QUESTION – THREE	73
QUESTION - FOUR	74
Answer Sheet – Exercise 14	75
EXERCISE 15: Multi Part Numerical Reasoning	79
QUESTION – ONE	79
QUESTION – TWO	81
QUESTION – THREE	82
QUESTION – FOUR	83
Answer Sheet – Exercise 15	85
EXERCISE 16: Numerical Reasoning – Answer Matching	89
EXERCISE 17: Numerical Reasoning – Answer Matching	93
EXERCISE 18: Numerical Reasoning – Answer Matching	96
EXERCISE 19: Numerical Reasoning – Answer Matching	99
EXERCISE 20: Numerical Reasoning – Answer Matching	102
ANSWERS:	106

SHORT MATHS
NUMERICAL REASONING

EXERCISE 1: Short Numerical Reasoning

Instructions: Work out and select ONE correct answer for each of the questions. Mark your answer on the answer sheet. *Maximum time allowed: 10 minutes*

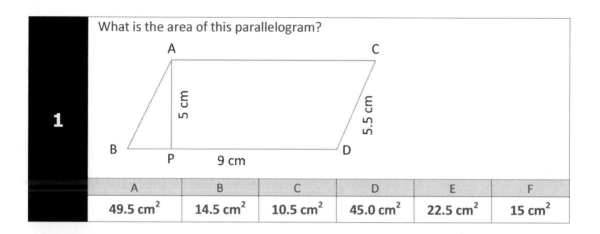

1 — What is the area of this parallelogram?

A	B	C	D	E	F
49.5 cm²	14.5 cm²	10.5 cm²	45.0 cm²	22.5 cm²	15 cm²

2 — Sally works 40 hours a week and her weekly wages is £1320. What is the hourly wage rate?

A	B	C	D	E	F
£189	£330	£33	£176	£182	£301

3 — Simon bought a car for £7500. After two years he sold it for £5000. What is the loss he made as a percentage?

A	B	C	D	E	F
25%	15%	0.33%	33%	0.25%	0.15%

4 — A train from Waterloo travels for 5 hours and 28 minutes to reach Edinburgh at 12:15 PM. What time did the train leave Waterloo?

A	B	C	D	E	F
06:47 AM	06:47 PM	06:57 AM	07:07 AM	05:47 AM	07:47 AM

5	Find the value of the following equation, where x is 3 and y is 1. $$\sqrt{2x^3 - 5y^2}$$					
	A	B	C	D	E	F
	13	1	49	7	8	64

6	What is 64% of £250?					
	A	B	C	D	E	F
	160	£80	£160	£16	£84	£25

7	How many 250 ml bottles can be filled with 15000 cm^3 of some liquid?					
	A	B	C	D	E	F
	60	6	600	15	20	25

8	A printer can print a maximum of 5 pages per minute. How many minutes would it take to print a document of 250 pages?					
	A	B	C	D	E	F
	50	5	10	15	18	21

9	What is $\sqrt{\left(2\frac{7}{8}\right)^2}$?					
	A	B	C	D	E	F
	$\frac{23}{8}$	$\frac{23}{16}$	$2\frac{49}{64}$	$4\frac{49}{64}$	$\frac{3}{8}$	$\frac{7}{8}$

10	What is $\dfrac{3.5 \times 3 + 2.5 \times 3}{4 \times 0 + 1.5 \times 2}$?					
	A	B	C	D	E	F
	3	6	12	2.57	5.14	1

11	How many rectangular tiles of 50 cm^2 are required to cover a surface area of 10 m^2?					
	A	B	C	D	E	F
	20	200	2000	50	500	5000

12	What is the missing number in this equation? $140 \times 2 = \boxed{} \times 4$					
	A	B	C	D	E	F
	210	220	70	280	1120	2240

13	80 children and 5 teachers are going on a school trip. A minibus can carry 25 people. How many minibuses are required?					
	A	B	C	D	E	F
	1	2	3	4	5	6

14	What is the product of 2.5 and 20?					
	A	B	C	D	E	F
	4	4.5	5	50	500	5.5

15	How many faces does a cube have?					
	A	B	C	D	E	F
	3	4	6	8	12	18

Answer Sheet:

1	A	B	C	D	E	F
2	A	B	C	D	E	F
3	A	B	C	D	E	F
4	A	B	C	D	E	F
5	A	B	C	D	E	F
6	A	B	C	D	E	F
7	A	B	C	D	E	F
8	A	B	C	D	E	F
9	A	B	C	D	E	F
10	A	B	C	D	E	F
11	A	B	C	D	E	F
12	A	B	C	D	E	F
13	A	B	C	D	E	F
14	A	B	C	D	E	F
15	A	B	C	D	E	F

EXERCISE 2: Short Numerical Reasoning

Instructions: Work out and select ONE correct answer for each of the questions. Mark your answer on the answer sheet. *Maximum time allowed: 10 minutes*

1	How many times does 0.25 go into 125?					
	A	B	C	D	E	F
	50	100	125	150	250	500

2	What is the value of "x" in the following equation? $$2x^2 + 5 = 205$$					
	A	B	C	D	E	F
	3	6	9	10	15	20

3	The temperature on a winter night is 2°C. What is the temperature if it drops further by a 5°C?					
	A	B	C	D	E	F
	7°C	3°C	-3°C	-2°C	-1°C	-5°C

4	What is the area of a square whose perimeter equates to 36 cm?					
	A	B	C	D	E	F
	9 cm	18 cm^2	81 cm^2	9 cm^2	324 cm^2	162 cm^2

5	The distance between two towns is 180 miles. If a car travels at an average speed of 45 miles per hour, then how long would the journey take?					
	A	B	C	D	E	F
	4 minutes	40 minutes	4 hours	6 hours	4.5 hours	25 minutes

6	The time in Alaska is 9 hours behind that of London. Samuel is flying from London to Alaska. His flight departs at 10:35 AM from London. The journey takes 9 hours and 30 minutes. What time will he arrive in Alaska?					
	A	B	C	D	E	F
	11:05 AM	11:05 PM	8:05 PM	8:05 AM	11:10 AM	12:05 PM

7	There are 15 large containers with each holding 25 boxes of dozen eggs. How many eggs are there in total?					
	A	B	C	D	E	F
	375	2250	4500	9000	6750	9375

8	What is $\frac{1}{3}^{rd}$ of 1.5 litres in ml?					
	A	B	C	D	E	F
	0.5 ml	50 ml	500 ml	0.5 litre	250 ml	1 litre

9	Which of the following is the same as $\frac{7}{15}$?					
	A	B	C	D	E	F
	0.36	0.467	0.466	0.366	2.14	2.142

10	Tickets to a play costs £7.50 for an adult and £3.50 for a child. What is the total cost if a couple took their two children to see the play?					
	A	B	C	D	E	F
	£37.00	£22.00	£14.50	£15.00	£33.50	£31.50

11	Victoria's score in an exam are as below: English – 89 out of 100 Maths – 91 out of 100 Science – 78 out of 100 What is the average percentage score for Victoria?

	A	B	C	D	E	F
	85%	85.6%	86%	87%	88%	91%

12	Which of the following is the 10th prime number?

	A	B	C	D	E	F
	17	19	23	29	31	37

13	Find the volume of a cube of length 6 cm?

	A	B	C	D	E	F
	64 cm^3	128 cm^3	216 cm^3	512 cm^3	36 cm^3	42 cm^3

14	What is the radius of a circle whose diameter is 7.5 cm?

	A	B	C	D	E	F
	7.5 cm	15 cm	10 cm	3.75 cm	3.70 cm	7.6 cm

15	Find the value of $\left(2\frac{3}{4}\right)^2$

	A	B	C	D	E	F
	$\frac{121}{16}$	$4\frac{9}{16}$	$2\frac{9}{16}$	$4\frac{6}{8}$	$2\frac{6}{8}$	$\frac{11}{16}$

Answer Sheet:

1	A	B	C	D	E	F
2	A	B	C	D	E	F
3	A	B	C	D	E	F
4	A	B	C	D	E	F
5	A	B	C	D	E	F
6	A	B	C	D	E	F
7	A	B	C	D	E	F
8	A	B	C	D	E	F
9	A	B	C	D	E	F
10	A	B	C	D	E	F
11	A	B	C	D	E	F
12	A	B	C	D	E	F
13	A	B	C	D	E	F
14	A	B	C	D	E	F
15	A	B	C	D	E	F

EXERCISE 3: Short Numerical Reasoning

Instructions: Work out and select ONE correct answer for each of the questions. Mark your answer on the answer sheet. *Maximum time allowed: 10 minutes*

1	What is **2 x 32 / (10 + 6) ?**					
	A	B	C	D	E	F
	12.4	10.4	4.0	2.0	11.4	12.8

2	A train travels at a speed of 85 miles per hour, for 3.5 hours. What is the total distance travelled in miles?					
	A	B	C	D	E	F
	297.5	320	287.5	282.5	280.0	295.5

3	What is **5% of 40% of 50% of 300?**					
	A	B	C	D	E	F
	30	3	4.5	13	1	33

4	Roman goes shopping and spends £180. He qualifies for a special discount of 20% on his total spending, as he has spent more than £100. How much will Roman have to pay after the discount?					
	A	B	C	D	E	F
	144	164	180	174	154	36

5	Prize money of £150,300 was shared by 15 people equally. How much money would each person receive?					
	A	B	C	D	E	F
	£10,020	£10.02	£1020	£1022	£1021	£120

6	Shakespeare is believed to be born in the year 1564 and died in the year 1616. How old was he when he died?					
	A	B	C	D	E	F
	62	42	52	53	51	50

7	At a stationery shop, a pencil costs £0.39 and a pen costs £0.65. Sarah buys 15 pencils and 10 pens. What is the total cost to Sarah?					
	A	B	C	D	E	F
	£123.50	£12.35	£10.40	£15.60	£14.20	£142.00

8	Find the area of the triangle ABC.					

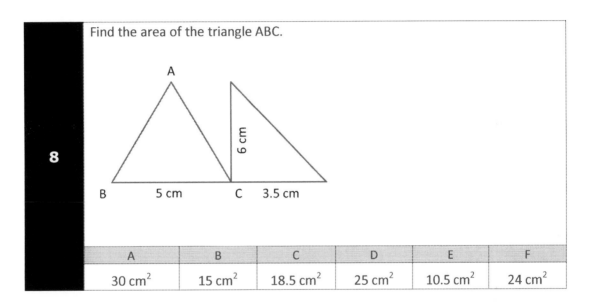

	A	B	C	D	E	F
	30 cm^2	15 cm^2	18.5 cm^2	25 cm^2	10.5 cm^2	24 cm^2

9	What is square root of 225?					
	A	B	C	D	E	F
	25	5	15	13.5	15.5	20

10	Which of the following is a triangular number?					
	A	B	C	D	E	F
	4	7	9	14	28	2

11	What is $^2/_5{}^{th}$ of 300?					
	A	B	C	D	E	F
	750	120	25	125	225	130

12	What the perimeter of a square whose area is 196 cm^2?					
	A	B	C	D	E	F
	5.6 cm	56 mm	56 cm	56 cm^2	5.6 cm^2	560 mm^2

13	Michael can run 3 laps in 5 minutes. How many **full** laps can he run in 8 minutes?					
	A	B	C	D	E	F
	4	5	4.8	15	4.2	4.6

14	How many Vertices does a cube have?					
	A	B	C	D	E	F
	4	6	8	10	12	2

15	Find the value of the following equation, where x is 3 and y is 5. $$2x + 5y^2$$					
	A	B	C	D	E	F
	131	125	55	45	135	51

Answer Sheet:

1	A	B	C	D	E	F
2	A	B	C	D	E	F
3	A	B	C	D	E	F
4	A	B	C	D	E	F
5	A	B	C	D	E	F
6	A	B	C	D	E	F
7	A	B	C	D	E	F
8	A	B	C	D	E	F
9	A	B	C	D	E	F
10	A	B	C	D	E	F
11	A	B	C	D	E	F
12	A	B	C	D	E	F
13	A	B	C	D	E	F
14	A	B	C	D	E	F
15	A	B	C	D	E	F

EXERCISE 4: Short Numerical Reasoning

Instructions: **Work out and select ONE correct answer for each of the questions. Mark your answer on the answer sheet.** *Maximum time allowed: 10 minutes*

1

What is the next number in the sequence below?

1   4   9   16   25   36

A	B	C	D	E	F
40	45	46	47	48	49

2

Gabriel has fiction and non-fiction books in the ratio of 3:4 respectively. If he has 140 books in total, how many of them are non-fiction?

A	B	C	D	E	F
30	60	80	100	40	35

3

What is $\dfrac{\frac{7}{8}}{\frac{8}{7}}$?

A	B	C	D	E	F
1	$^{49}/_{64}$	$^{64}/_{49}$	49	64	$^{3}/_{6}$

4

What do **1.75 litres + 300 ml** equate to?

A	B	C	D	E	F
301.75 litres	1.05 litres	1.35 litres	2.05 litres	1.78 litres	4.75 litres

5

What is 25% of 8 km?

A	B	C	D	E	F
2 miles	4 km	4.5 km	6 km	2.5 km	2 km

6

What is the area of this shape?

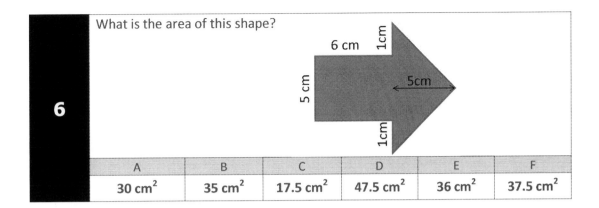

A	B	C	D	E	F
30 cm²	35 cm²	17.5 cm²	47.5 cm²	36 cm²	37.5 cm²

7

Find the value of the expression below, if a = 5.

$$\frac{a^2 + 3a}{2a}$$

A	B	C	D	E	F
4	35	25	5	10	40

8

Find the angle marked p° below?

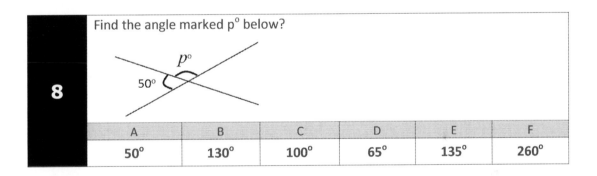

A	B	C	D	E	F
50°	130°	100°	65°	135°	260°

9

What is the product of 1.53 and 100?

A	B	C	D	E	F
1.53	15.3	153	1530	101.53	10.153

10

How many edges does a triangular prism have?

A	B	C	D	E	F
3	6	8	9	12	16

11	Which one of the below is a square number?					
	A	B	C	D	E	F
	2	3	5	10	64	122

12	What is $9^2 + \sqrt{36}$?					
	A	B	C	D	E	F
	87	88	89	90	14	15

13	Rob is travelling to New York and wants to buy some US Dollars. The exchange rate is 1.6 dollars to a pound. How many US Dollars will he receive for £150?					
	A	B	C	D	E	F
	$240	$93.75	$180	$200	$160	$190

14	Miranda spends 6 hours at school each working day. What is this as a fraction of the day?					
	A	B	C	D	E	F
	$^1/_6$	$^1/_4$	$^2/_3$	$^3/_4$	$^1/_5$	$^1/_2$

15	Diego is a striker for a local football club and his average goals scored per game is 0.6. He has played a total of 30 games. How many goals has he scored?					
	A	B	C	D	E	F
	10	12	14	16	18	22

Answer Sheet:

	A	B	C	D	E	F
1	A	B	C	D	E	F
2	A	B	C	D	E	F
3	A	B	C	D	E	F
4	A	B	C	D	E	F
5	A	B	C	D	E	F
6	A	B	C	D	E	F
7	A	B	C	D	E	F
8	A	B	C	D	E	F
9	A	B	C	D	E	F
10	A	B	C	D	E	F
11	A	B	C	D	E	F
12	A	B	C	D	E	F
13	A	B	C	D	E	F
14	A	B	C	D	E	F
15	A	B	C	D	E	F

EXERCISE 5: Short Numerical Reasoning

Instructions: **Work out and select ONE correct answer for each of the questions. Mark your answer on the answer sheet.** *Maximum time allowed: 10 minutes*

1

What is the next number in the sequence below?

3.07   3.14   3.21   3.28   3.35   ?

A	B	C	D	E	F
3.44	3.41	3.42	3.43	3.40	3.37

2

What is 3.15 divided by 100?

A	B	C	D	E	F
31.5	0.0315	0.315	3.15	0.00315	315

3

Matthew gets a letter a letter from his employer informing him that his annual salary will be increased from next month by 10% to £82,500. What is his current annual salary?

A	B	C	D	E	F
£90,750	£80,000	£89,000	£78,000	£75,000	£67,000

4

What is $^2/_3{}^{rd}$ of a litre in millilitres?

A	B	C	D	E	F
0.66 ml	6.66 ml	66.6 ml	666.67 ml	0.67 l	233 ml

5

What is the square root of 400?

A	B	C	D	E	F
40	20	10	8	4	2

6	Which of the following fractions represent 80%?					
	A	B	C	D	E	F
	$^1/_8$	$^2/_4$	$^3/_4$	$^4/_5$	$^5/_7$	$^6/_8$

7	A car can travel 8 km per litre of petrol. Its tank when full holds 60 litres of petrol. What is the distance it can travel on a half the tank of petrol?					
	A	B	C	D	E	F
	48 km	48 miles	480 km	24 km	240 km	320 km

8	What is the maximum individual angle possible inside an equilateral triangle?					
	A	B	C	D	E	F
	180°	145°	90°	60°	45°	30°

9	Find the area of the shaded part of the diagram below.					

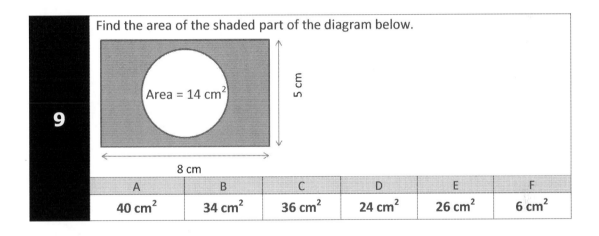

	A	B	C	D	E	F
	40 cm^2	34 cm^2	36 cm^2	24 cm^2	26 cm^2	6 cm^2

10	How many vertices does a cuboid have?					
	A	B	C	D	E	F
	12	6	8	4	15	16

11	What is $\sqrt{(12 \times 0.5)^2 + (4 / 0.5)^2}$?					
	A	B	C	D	E	F
	10	42	25	36	5	15

12	Three copies of a new music album are being sold at a local shop every 10 minutes. How many copies will sell in 3 hours?					
	A	B	C	D	E	F
	24	34	44	54	64	12

13	At a wedding dinner 32 bottles of cola were consumed. Each bottle contained 1.5 litres of cola. How many litres of cola is that in total?					
	A	B	C	D	E	F
	48 l	4.8 l	96 l	26 l	12 l	24 l

14	What is the remainder when you divide 20 by 6?					
	A	B	C	D	E	F
	3.33	3.34	2	18	6.33	$3\,^2/_6$

15	What is $^1/_6{}^{th}$ of a kilometre in metres?					
	A	B	C	D	E	F
	0.66 m	6.66 m	666 m	16.6 m	166.6 m	166.67 m

Answer Sheet:

1	A	B	C	D	E	F
2	A	B	C	D	E	F
3	A	B	C	D	E	F
4	A	B	C	D	E	F
5	A	B	C	D	E	F
6	A	B	C	D	E	F
7	A	B	C	D	E	F
8	A	B	C	D	E	F
9	A	B	C	D	E	F
10	A	B	C	D	E	F
11	A	B	C	D	E	F
12	A	B	C	D	E	F
13	A	B	C	D	E	F
14	A	B	C	D	E	F
15	A	B	C	D	E	F

EXERCISE 6: Short Numerical Reasoning

Instructions: Work out and select ONE correct answer for each of the questions. Mark your answer on the answer sheet. *Maximum time allowed: 10 minutes*

1 — If 30th of October is a Thursday, what day is 4th November?

A	B	C	D	E	F
Monday	Tuesday	Wednesday	Thursday	Friday	Saturday

2 — What is $\sqrt{(3.15 + 0.85)^2}$?

A	B	C	D	E	F
2	4	4.15	6	12	10.5

3 — The time zone in India is 5:30 hours ahead that of London. If it is 10:15 AM in India, what is the time in London?

A	B	C	D	E	F
5:45AM	4:45 AM	4:30 AM	5:30 AM	5:00 AM	6:15 AM

4 — What is the perimeter of a square whose area is 36 cm^2?

A	B	C	D	E	F
12 cm	24 cm	18 cm	30 cm	36 cm	42 cm

5 — The distance between London and Birmingham is 120 miles. If a train has completed 60% of the journey, how many miles of the journey is remaining?

A	B	C	D	E	F
72 miles	82 miles	78 miles	48 miles	58 miles	62 miles

6

One in every five of the 200 bottles of orange juice has been sold in a shop. How many are remaining?

A	B	C	D	E	F
40	60	80	120	160	180

7

How many hours make 5 days?

A	B	C	D	E	F
60	120	240	300	360	400

8

The circumference of a bike's wheel is 2 metres. How many times will the wheel rotate if the bike travelled a distance of 2 kilometres?

A	B	C	D	E	F
10	100	1000	20	200	2000

9

A bag of oranges weigh 2.1 kg. If the average weight of an orange is 300 grams, how many oranges are there in the bag?

A	B	C	D	E	F
4	5	6	7	8	9

10

How many faces does an hexagonal prism has?

A	B	C	D	E	F
4	6	8	12	16	20

11

Find the square root of 256.

A	B	C	D	E	F
14	12	13	15	16	25

12	A gallon is approximately 4.5 litres. A car's petrol tank holds 60 litres of petrol. Rounded to the nearest tenth of a gallon, how many gallons is that?					
	A	B	C	D	E	F
	13.3	13.4	13.33	13.34	13	14

13	Find the angle marked m° 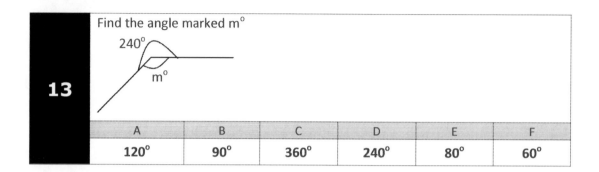					
	A	B	C	D	E	F
	120°	90°	360°	240°	80°	60°

14	How many 0.3s are there in 45?					
	A	B	C	D	E	F
	15	150	30	350	40	450

15	To get a particular shade of colour, Alexandra is mixing 5 parts of red paint with 3 parts blue. If she needs 40 litres of mixed paint, how many litres of red paint is required?					
	A	B	C	D	E	F
	25 l	15 l	5 l	32 l	21 l	33 l

Answer Sheet:

1	A	B	C	D	E	F
2	A	B	C	D	E	F
3	A	B	C	D	E	F
4	A	B	C	D	E	F
5	A	B	C	D	E	F
6	A	B	C	D	E	F
7	A	B	C	D	E	F
8	A	B	C	D	E	F
9	A	B	C	D	E	F
10	A	B	C	D	E	F
11	A	B	C	D	E	F
12	A	B	C	D	E	F
13	A	B	C	D	E	F
14	A	B	C	D	E	F
15	A	B	C	D	E	F

EXERCISE 7: Short Numerical Reasoning

Instructions: Work out and select ONE correct answer for each of the questions. Mark your answer on the answer sheet. *Maximum time allowed: 10 minutes*

1	The lowest temperature last night was -4°C. What is the temperature today if it falls by a further 5°C?					
	A	B	C	D	E	F
	1°C	-1°C	9°C	-9°C	7°C	11°C

2	A car is travelling at a speed of 50.25 miles per hour. How far it would travel in 40 minutes?					
	A	B	C	D	E	F
	33.68 mph	45.25 miles	62.25 miles	70 miles	33.50 miles	40 miles

3	What is the next number in sequence? -3 -8 -14 -21 -29					
	A	B	C	D	E	F
	-38	-39	-40	-37	-41	-42

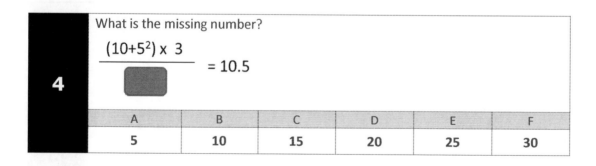

4	What is the missing number? $$\frac{(10+5^2) \times 3}{\boxed{}} = 10.5$$					
	A	B	C	D	E	F
	5	10	15	20	25	30

5	Which of the following is a prime number?					
	A	B	C	D	E	F
	1	4	6	8	11	28

6	Which of the following is a concave angle?					
	A	B	C	D	E	F
	90°	45°	125°	180°	175°	245°

7	How many cola cans of 300 ml capacity can be filled by 4.5 litres of cola?					
	A	B	C	D	E	F
	15	150	45	35	32	31

8	A room is 20 feet long and 10 feet wide. The total cost to fit wooden floor is £500. How much does the flooring cost per square feet?					
	A	B	C	D	E	F
	£25	£2.50	£12.50	£50	£32.50	£11.25

9	The area of a square is 625 cm^2. What is the length of the square?					
	A	B	C	D	E	F
	100 cm	25 cm	50 cm	75 cm	125 cm	10.25 cm

10	Danish is saving all his pocket money to buy the full football kit of his favourite club, which costs £75. His pocket money is £15 pounds every two weeks. How long would it take for Danish to save enough to buy the kit?					
	A	B	C	D	E	F
	5 weeks	8 weeks	10 weeks	12 weeks	15 weeks	20 weeks

11	Paul decides to donate 5% of his salary this month to a charity. His annual salary is £60,000. How much does he give to charity?					
	A	B	C	D	E	F
	£3000	£416.67	£500	£300	£400	£250

12	At a popular Grammar school, only one child in every five passes the entrance exam. If there are 120 seats available, how many children normally take the test?					
	A	B	C	D	E	F
	600	500	580	590	575	625

13	How many faces does a Sphere have?					
	A	B	C	D	E	F
	0	1	2	4	6	8

14	You are asked to think of a number, double it and then add 7. Divide the whole thing by 10. Which of the following equations represent this?					
	A	B	C	D	E	F
	$n+n+7/10$	$2n+7/10$	$n^2+7/10$	$(n^2+7)/10$	$(2n+7)/10$	$2(n+7/10)$

15	A family spends £900 every month on house rent. The monthly income of that family is £3600. What fraction of the salary does the rent represent?					
	A	B	C	D	E	F
	$\frac{1}{2}$	$\frac{3}{2}$	$\frac{5}{2}$	$\frac{1}{3}$	$\frac{1}{4}$	$\frac{1}{5}$

Answer Sheet:

1	A	B	C	D	E	F
2	A	B	C	D	E	F
3	A	B	C	D	E	F
4	A	B	C	D	E	F
5	A	B	C	D	E	F
6	A	B	C	D	E	F
7	A	B	C	D	E	F
8	A	B	C	D	E	F
9	A	B	C	D	E	F
10	A	B	C	D	E	F
11	A	B	C	D	E	F
12	A	B	C	D	E	F
13	A	B	C	D	E	F
14	A	B	C	D	E	F
15	A	B	C	D	E	F

EXERCISE 8: Short Numerical Reasoning

Instructions: Work out and select ONE correct answer for each of the questions. Mark your answer on the answer sheet. *Maximum time allowed: 10 minutes*

1 Provide the total of the following.

0.329 + 0.021 + 0.212

A	B	C	D	E	F
0.56	0.562	0.751	0.741	0.742	0.57

2 The perimeter of a square is 36 cm. What is the area?

A	B	C	D	E	F
91 cm^2	18 cm^2	36 cm^2	81 cm^2	92 cm^2	42 cm^2

3 A car is travelling at 35 metres per second. What is the distance in kilometres if it the car travelled for 2 hours at that speed?

A	B	C	D	E	F
25.2 km	252 km	52.2 km	420 km	4.2 km	424 km

4 A shop only sells eggs in boxes of 12. Raymond needs 50 eggs. How many boxes will he need to buy?

A	B	C	D	E	F
4	5	6	3	7	8

5 Take away 5% of £3000 from £4000.

A	B	C	D	E	F
£1000	£2500	£3850	£2850	£3150	£150

6	What is the volume of a cube of length 5 cm?					
	A	B	C	D	E	F
	25 cm^2	125 cm^2	125cm^3	625 cm^3	255 cm^3	150 cm^3

7	Which of the following is NOT a triangular number?					
	A	B	C	D	E	F
	3	6	10	14	15	28

8	What is the angle represented by letter m in the diagram below.

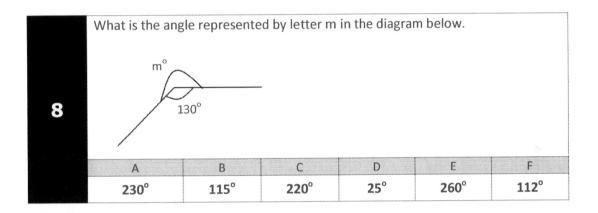

	A	B	C	D	E	F
	230°	115°	220°	25°	260°	112°

9	What is the next fraction in the sequence?

$^1/_{10}$ $^3/_{20}$ $^2/_{10}$ $^1/_4$ $^3/_{10}$

A	B	C	D	E	F
$^{13}/_{10}$	$^2/_5$	$^7/_{10}$	$^7/_{20}$	$^9/_{10}$	$^3/_{10}$

10	300 grams of sugar is required to make a cake weighing 2 kg. How much sugar is required to make a three-tier cake weighing 5 kg?					
	A	B	C	D	E	F
	650 g	700 g	750 g	0.5 kg	1.25 kg	800 g

11	Which of the following shapes have just ONE edge?					
	A	B	C	D	E	F
	Cube	Cylinder	Sphere	Cone	Triangular Prism	Hexagonal Prism

12	Which of the following percentages of "a" does the value of "y" represent in the following equation? $0.1\,a = y$					
	A	B	C	D	E	F
	1%	10%	25%	33%	66%	72%

13	What is the order of the rotational symmetry of the following shape?					

	A	B	C	D	E	F
	0	1	2	3	4	8

14	Anthony's car travel 9 km on a litre of petrol and the petrol tank's maximum capacity is 50 litres. How many times does the petrol needs refilling for a journey of 1250 km?					
	A	B	C	D	E	F
	450	45	2	3	1	4

15	Wayne has saved £125. He spends $^3/_5$th of his money on shopping. How much money is left?					
	A	B	C	D	E	F
	£75	£50	£60	£45	£40.50	£90

Answer Sheet:

	A	B	C	D	E	F
1	A	B	C	D	E	F
2	A	B	C	D	E	F
3	A	B	C	D	E	F
4	A	B	C	D	E	F
5	A	B	C	D	E	F
6	A	B	C	D	E	F
7	A	B	C	D	E	F
8	A	B	C	D	E	F
9	A	B	C	D	E	F
10	A	B	C	D	E	F
11	A	B	C	D	E	F
12	A	B	C	D	E	F
13	A	B	C	D	E	F
14	A	B	C	D	E	F
15	A	B	C	D	E	F

EXERCISE 9: Short Numerical Reasoning

Instructions: Work out and select ONE correct answer for each of the questions. Mark your answer on the answer sheet. *Maximum time allowed: 10 minutes*

1	Today 1 pound is equal to 1.3 Euros. How many Euros will you get for £200?					
	A	B	C	D	E	F
	£260	€200	€130	€154	€210	€250

2	Find the next number in the sequence?					
	3 6 10 15 21 27					
	A	B	C	D	E	F
	28	36	37	35	33	42

3	Which of the following shapes have no vertices?					
	A	B	C	D	E	F
	Cone	Triangular Prism	Cuboid	Cube	Octahedron	None of these

4	Find the square root of 1225.					
	A	B	C	D	E	F
	15	25	35	45	55	65

5	The width of a rectangle is 5 cm and the length is 1.5 cm longer than the width. What is the perimeter of the rectangle?					
	A	B	C	D	E	F
	22 cm	23 cm	21 cm	11.5 cm	12.5 cm	13 cm

| 6 | What is the angle marked "m" in the diagram below? |

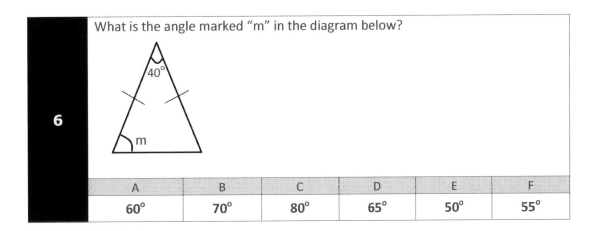

	A	B	C	D	E	F
	60°	70°	80°	65°	50°	55°

| 7 | What is the decimal equivalent of $3\dfrac{2}{100}$? |

	A	B	C	D	E	F
	3.2	3.22	3.02	3.12	30.2	103.2

| 8 | Evaluate: $\sqrt{(5.15 - 15/100)^2}$ |

	A	B	C	D	E	F
	5	25	5.25	6.25	62.5	625

| 9 | A bus journey to school takes 23 minutes. The bus reaches school at 08:42 AM. What did it start its journey? |

	A	B	C	D	E	F
	8:21 AM	8:20 AM	8:19 PM	8:19 AM	8:29 AM	8:22 AM

| 10 | Reduce 3.5 litre by 20% |

	A	B	C	D	E	F
	0.7 l	700 ml	2.75 l	2.8 l	2.9l	2.85 l

11 How many 20p coins are required to make £200?

A	B	C	D	E	F
10	100	200	1000	150	250

12 The area of the triangle below is 21 cm^2. Find the height.

6cm

A	B	C	D	E	F
3.5 cm	7 cm	5 cm	6 cm	5.5 cm	4.2 cm

13 Consider the equation $r = d - 30$

Which of the options is the same as the above equation?

A	B	C	D	E	F
d = r − 30	d = 30 − r	30 = r − d	r + 30 = d	r/d = -30	d/r = -30

14 It is believed that a family of four generates a rubbish of 1000 kg each year. How much rubbish would a family of 5 would generate in 6 months?

A	B	C	D	E	F
625 kg	250 kg	1000 kg	500 kg	750 kg	650 kg

15 Find the missing number.

$$\frac{(6^2 + 8^2) \times 4}{\boxed{}} = 20$$

A	B	C	D	E	F
20	10	200	40	15	100

Answer Sheet:

1	A	B	C	D	E	F
2	A	B	C	D	E	F
3	A	B	C	D	E	F
4	A	B	C	D	E	F
5	A	B	C	D	E	F
6	A	B	C	D	E	F
7	A	B	C	D	E	F
8	A	B	C	D	E	F
9	A	B	C	D	E	F
10	A	B	C	D	E	F
11	A	B	C	D	E	F
12	A	B	C	D	E	F
13	A	B	C	D	E	F
14	A	B	C	D	E	F
15	A	B	C	D	E	F

EXERCISE 10: Short Numerical Reasoning

Instructions: Work out and select ONE correct answer for each of the questions. Mark your answer on the answer sheet. *Maximum time allowed: 10 minutes*

1

The area of a triangle is calculated using the formula below:

Area = 2 x 3.14 x radius

Using the formula above calculate the area of the circle below.

A	B	C	D	E	F
37.68 cm^2	18.84 cm^2	9.42 cm^2	28.26 cm^2	24 cm^2	32 cm^2

2

A large hose is pumping 2.5 litres of water every 30 seconds. How much water will flow through in 30 minutes?

A	B	C	D	E	F
50 litres	150 litres	100 litres	75 litres	65 litres	165 litres

3

One gallon is the same is 4.54 litres.
How many gallons of petrol are required to fill a tank with a capacity of 45.4 litres?

A	B	C	D	E	F
200	20	9.68	2	11.76	10

4

Sam's car does 10 km for a litre of petrol. The cost per litre of petrol is £1.35. What is the cost of petrol for journey of 450 km?

A	B	C	D	E	F
£607.50	£67.50	£135	£60.75	£60.25	£55.25

5 The price of a house in London has increased every year by 5% over the last three years. If Sanjay bought a house for £210,000 three years ago, what is the price of his house now? Round the answer to the nearest pound.

A	B	C	D	E	F
£241,500	£220,500	£243,101	£221,101	£252,101	£252,000

6 Find a number that is half way between 20.4 and 24.

A	B	C	D	E	F
22.2	22.1	22.3	22.4	21.2	21.3

7 What is $\frac{1}{5}^{th}$ of 750?

A	B	C	D	E	F
15	150	125	130	135	120

8 Find the next number in the sequence.

1   8   9   64   25   216   49   512

A	B	C	D	E	F
58	64	1032	1024	46	81

9 What is the cube root of 64?

A	B	C	D	E	F
8	7	6	4	5	3

10 What is the angle between two hands of a clock at 3:00 AM?

A	B	C	D	E	F
45°	90°	120°	150°	180°	225°

11	Find the angle "y" in the diagram below.

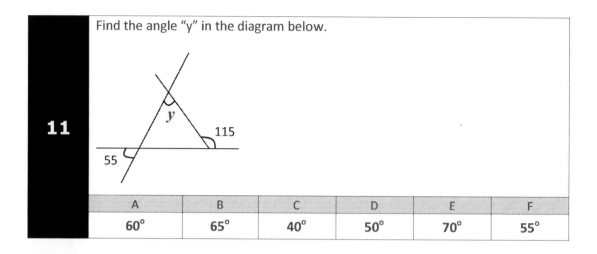

A	B	C	D	E	F
60°	65°	40°	50°	70°	55°

12	There are 365 days in a non-leap year. How many full weeks are there in a year?

A	B	C	D	E	F
51	52	53	49	50	52.14

13	Find the height of a cuboid with a base measuring 7 cm by 9 cm and a volume of $315cm^3$

4 cm	3 cm	5 cm	7 cm	8 cm	10 cm

14	Find the missing number.

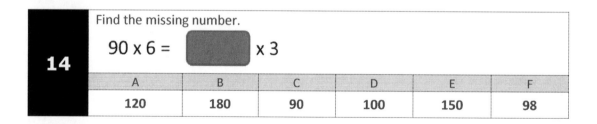

A	B	C	D	E	F
120	180	90	100	150	98

15	What is the product of 0.00321 and 1000?

A	B	C	D	E	F
0.321	3.21	32.10	321.0	0.0321	3210

Answer Sheet:

1	A	B	C	D	E	F
2	A	B	C	D	E	F
3	A	B	C	D	E	F
4	A	B	C	D	E	F
5	A	B	C	D	E	F
6	A	B	C	D	E	F
7	A	B	C	D	E	F
8	A	B	C	D	E	F
9	A	B	C	D	E	F
10	A	B	C	D	E	F
11	A	B	C	D	E	F
12	A	B	C	D	E	F
13	A	B	C	D	E	F
14	A	B	C	D	E	F
15	A	B	C	D	E	F

LONG MATHS
NUMERICAL REASONING

EXERCISE 11: Multi Part Numerical Reasoning

Instructions: Work out and mark your answers by filling the appropriate grids on the answer sheet. Use a separate sheet for your workings. *Maximum time allowed: 15 minutes*

QUESTION - ONE

Luke earns an annual salary of £90,000 as an IT manager.

The tax on this income is charged as per the rules below:
- *No tax on the first £10,000*
- *20% tax on income between £10,001 and £40,000*
- *40% tax on income above £40,000*

11.1 How much tax will Luke pay in a year? Round it to the nearest ten.

11.2 How much money does Luke gets paid after tax, every month? Round it to the nearest ten.

11.3 If Luke also received a bonus of £18,000, how much more tax he will have to pay that year? Round it to the nearest ten.

11.4 Express the bonus as a percentage of his basic salary. Round it to the nearest unit.

11.5 Luke is expecting his salary to be increased by 15% next year. What is the total tax he will pay next year? Round it to the nearest ten.

The pie chart below shows the favourite sports of children at a local school.

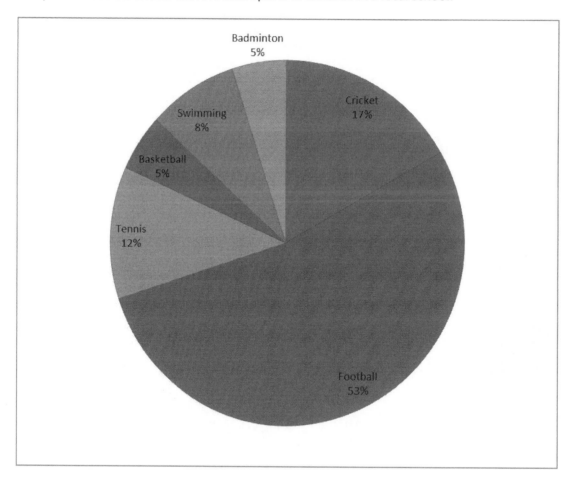

Answer the questions on the next page based on the data visualised in this pie chart.

11.6 Number of children for whom Football is the favourite sport is 160.
What is the total number of children in the school?

11.7 How many more children like Football than Cricket?

11.8 What is the number of children whose favourite sport is not Football?

11.9 Children, whose favourite sport is Badminton changed their minds and chose Tennis instead. What is the percentage of children whose favourite sport is Tennis?

11.10 The ratio of boys to girls is 2:3.
How many girls are in the school?

The results of ten students taking a maths test are below. The maximum mark allowed is 80.

67	48	59	63	59	32	45	72	75	70

11.11   What is the average score of the maths test?

11.12   Find the mode.

11.13   What is the median?

11.14   What is the maximum score as a percentage, rounded to the nearest unit?

11.15   Find the range.

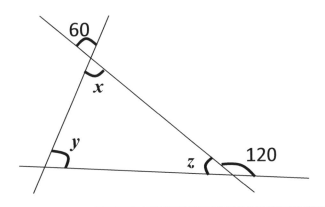

11.16 Find the value of the following equation?

$$x + y + z$$

11.17 Find the value of angle "x".

11.18 Find the value of angle "y".

11.19 Find the value of angle "z".

11.20 Using the values found in the previous steps, resolve the following equation.

$$2x + 3y / z$$

Answer Sheet – Exercise 11

Question Number	Answer Grid	Question Number	Answer Grid
11.1		**11.4**	
11.2		**11.5**	
11.3			

Answer Sheet – Exercise 11

Question Number	Answer Grid	Question Number	Answer Grid
11.6		**11.9**	
11.7		**11.10**	
11.8			

Answer Sheet – Exercise 11

Question Number	Answer Grid	Question Number	Answer Grid
11.11		**11.14**	
11.12		**11.15**	
11.13			

Answer Sheet – Exercise 11

Question Number	Answer Grid	Question Number	Answer Grid
11.16		**11.19**	
11.17		**11.20**	
11.18			

EXERCISE 12: Multi Part Numerical Reasoning

Instructions: Work out and mark your answers by filling the appropriate grids on the answer sheet. Use a separate sheet for your workings. *Maximum time allowed: 15 minutes*

QUESTION - ONE

A square of area 81 cm^2 is cut as shown below into two triangles.

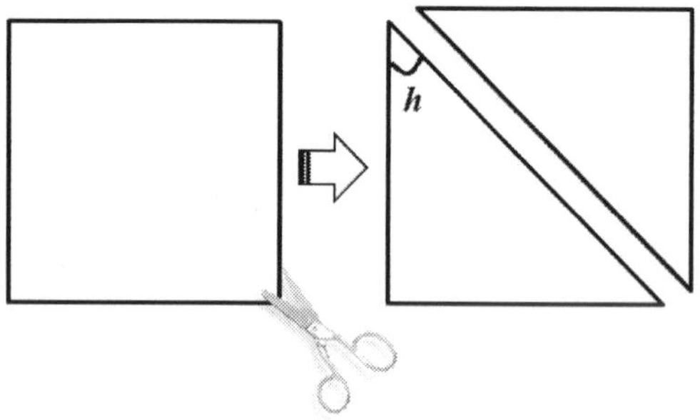

12.1   What is the length of each side of the square?

12.2   What is the area of each triangle, rounded to the nearest unit?

12.3   What is the maximum individual angle of the two triangles?

12.4   Find the value of angle "h".

12.5   What is the total of angles from both the triangles?

Jack won two hundred thousand pounds in lottery. He spent $\frac{2}{5}^{th}$ on paying towards buying a house. He spent £35,000 on buying a car and £15,000 on holidays. He also donated 2% of his winnings to a local charity.

12.6 How much did he pay towards buying a house?

12.7 How much did the local charity get?

12.8 What was the percentage of the winning money spent towards buying the house?

12.9 How much of the lottery prize money is left?

12.10 Jack deposits the remaining amount in a bank. He gets interest paid to him every year at the rate of 3%. How much interest does he accumulate in three years? Round it to the nearest unit.

Here are the GCSE results of a popular grammar school in Essex.

Subject	No. of Entries	No. of students by Levels achieved					
		A*	A	B	C	D	E
Art	22	13	6	2		1	
Biology	77	47	22	4	4		
Chemistry	78	52	18	8			
Computing	24	14	6	3	1		
Design & Technology	20	10	5	3	2		
English	99	71	15	9	3		1
French	99	62	14	8	12	3	
Geography	69	39	14	13	2	1	
Maths	99	78	21				
Physics	77	47	13	10			
Spanish	17	13	4				
Statistics	89	55	21	12	1		
Total	770	501	159	72	25	5	1

12.11 What is the total number of GCSE entries?

12.12 How many entries are there for languages, including English?

12.13 Physics, Chemistry and Biology are the three strands of Science. How many entries are there for Science in total?

12.14 What percentage of the total entries is for Science related subjects? Round it to the nearest unit.

12.15 What percentage of the entries achieved either an A* or A in total? Round it to the nearest unit.

 *London eye is one of the most popular attractions in London. The big wheel has 32 capsules and each capsule can carry 25 people in one go, providing spectacular views of London. Each rotation takes about 30 minutes to complete and the circumference of the wheel is 434 m.*

12.16 How many people can the wheel carry per revolution?

12.17 How many times does the wheel rotate in 8 hours?

12.18 Each capsule weights 10 tonnes. What is the total weight in tonnes of all the capsules?

12.19 What is the distance travelled by a capsule, in metre, for every revolution of the wheel?

12.20 Rob and his four friends pay £36 in total to get on London eye. What is the price per entry?

Answer Sheet – Exercise 12

Question Number	Answer Grid	Question Number	Answer Grid
12.1		**12.4**	
12.2		**12.5**	
12.3			

Answer Sheet – Exercise 12

Question Number	Answer Grid	Question Number	Answer Grid
12.6		**12.9**	
12.7		**12.10**	
12.8			

Answer Sheet – Exercise 12

Question Number	Answer Grid	Question Number	Answer Grid
12.11		**12.14**	
12.12		**12.15**	
12.13			

Answer Sheet – Exercise 12

Question Number	Answer Grid	Question Number	Answer Grid
12.16		**12.19**	
12.17		**12.20**	
12.18			

EXERCISE 13: Multi Part Numerical Reasoning

Instructions: Work out and mark your answers by filling the appropriate grids on the answer sheet. Use a separate sheet for your workings. *Maximum time allowed: 15 minutes*

The triangle below is a right angled triangle. The shaded rectangle fits inside the triangle perfectly.

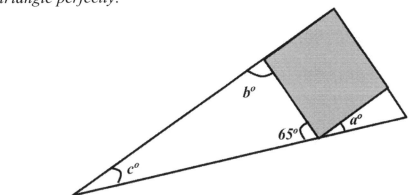

13.1	What is the value of angle "a"?

13.2	What is the value of angle "b"?

13.3	What is the value of angle "c"?

13.4	Find the value of the equation below: $a + (2b - 2c) / 2$

13.5	What is the value a^2?

Here is a shape formed of three rectangles. The length of the rectangle is twice the width. The outer perimeter of the combined shape is 28 cm.

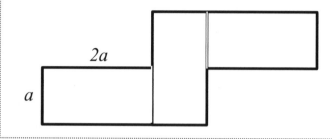

13.6 What is the value of "a" in cm?

13.7 What is the area in cm^2 of an individual rectangle?

13.8 What is the area of the overall shape formed?

13.9 What is the perimeter if the value of a is 4cm?

13.10 Find the value of the following equation.

$3(2a^2)$

The graph below shows the average high and low temperature recorded in Munich, Germany over a period of twelve months.

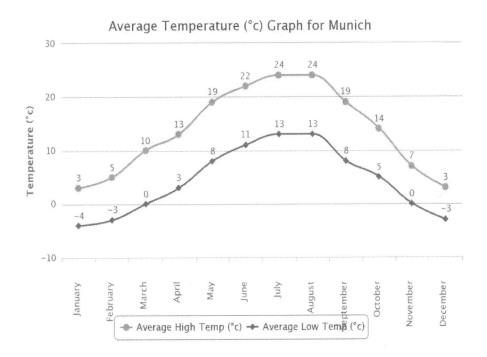

Based on the data presented in the graph, answer the following questions.

| 13.11 | What is the highest temperature recorded in the year? |

| 13.12 | What is the range of temperature? |

| 13.13 | What is the lowest (average) temperature recorded in the year? |

| 13.14 | What is the maximum difference between the highest and lowest recorded in the same month? |

| 13.15 | The lowest minimum temperature in Edinburgh is twice as low as the lowest average temperature recorded in Berlin. Note down the lowest temperature in Edinburgh. |

Jane and her husband Jack bought a home for £150,000. They paid 40% of that money using their savings. The remaining was paid for by a bank loan (mortgage). The bank would charge them 3.5% per annum in interest to be paid monthly.

They also need to pay the government a tax of 1% of the purchase price.

13.16   How much did Jack and Jane borrow from the bank?

13.17   How much of their savings did they use to purchase the house?

13.18   How much tax do they need pay the government? Round to the nearest pound.

13.19   What is the monthly interest payable to the bank?

13.20   The seller of the house needs to pay his agent a commission of 0.5%. How much money does the agent get paid?

Answer Sheet – Exercise 13

Question Number	Answer Grid	Question Number	Answer Grid
13.1		**13.4**	
13.2		**13.5**	
13.3			

Answer Sheet – Exercise 13

Question Number	Answer Grid	Question Number	Answer Grid
13.6		**13.9**	
13.7		**13.10**	
13.8			

Answer Sheet – Exercise 13

Question Number	Answer Grid	Question Number	Answer Grid
13.11		**13.14**	
13.12		**13.15**	
13.13			

Answer Sheet – Exercise 13

Question Number	Answer Grid	Question Number	Answer Grid
13.16		**13.19**	
13.17		**13.20**	
13.18			

EXERCISE 14: Multi Part Numerical Reasoning

Instructions: Work out and mark your answers by filling the appropriate grids on the answer sheet. Use a separate sheet for your workings. *Maximum time allowed: 15 minutes*

QUESTION – ONE

The pie chart below is the outcome of a survey to find the favourite way children prefer their potatoes cooked at a school in Kent. The total number of children who took part in the survey is 85.

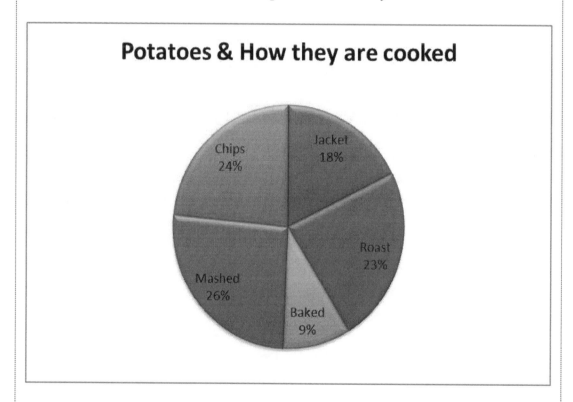

Potatoes & How they are cooked

Chips 24%
Jacket 18%
Roast 23%
Baked 9%
Mashed 26%

14.1   How many children like Mashed potatoes?

14.2   What is the total number of children who either like Chips or Roast potatoes?

14.3   A portion of Roast potatoes cost 25p. What is the cost of 20 portions?

14.4   10kg of potatoes are used every day in the kitchen. Applying the same percentages above, how much of this, in grams, is used for Mashed potatoes?

14.5   The school ran a healthy eating event, after which the number of children who liked Chips dropped by 30%. How many children is that?

Jerome, an accountant by profession, charges £300 a year for his services. He currently has 250 clients. The total annual expenses in providing those services, excluding his salary, are £15,000.

14.6   What is the annual income of Jerome for providing his accountancy services?

14.7   What percentage of the income are expenses?

14.8   Jerome would like to increase the revenue next year. One of the ways he is considering is to increase the charge rate to £325. How much more revenue would such a move generate?

14.9   Alternatively, he can run a campaign to sign up more clients. He expects to sign up 25 new clients. How much more revenue would this generate?

14.10 $^3/_4{}^{th}$ of the annual revenue is spent on wages. How much would this be?

QUESTION – THREE

Neil's mother is making cupcakes for a charity event. The following ingredients are required to make 25 cup cakes.

Butter	250 g	£2.50
Sugar	250 g	£2.50
Flour	250 g	£1.00
Eggs	4	£1.00
Other ingredients	N/A	£2.00

14.11 How much does it cost to make 25 cupcakes?

14.12 How much does it cost to make 250 cupcakes?

14.13 250 cupcakes were put on sale for 50p each at the charity event and were all sold out. How much profit would this make for the charity?

14.14 How many kilogram of sugar is required to make 500 cupcakes?

14.15 How many eggs are required to make 500 cupcakes?

QUESTION - FOUR

Raj is decorating his home and is getting a new carpet for his living room. The living room measures 50 feet in length and 30 feet in width.

14.16 How many square feet of carpet is required?

14.17 The price of square feet of carpet is £2.50. How much does it cost to carpet the living room?

14.18 He also wants to fix a skirting board around the borders of the living room wall. How many feet of skirting board is needed?

14.19 The skirting board is sold in lengths of one meter. 1 metre is equal to 3.28 feet. How many units of skirting boards are required?

14.20 The unit price of a metre skirting board is £5.00. What is the cost of skirting boards required?

Answer Sheet – Exercise 14

Question Number	Answer Grid	Question Number	Answer Grid
14.1		**14.4**	
14.2		**14.5**	
14.3			

Answer Sheet – Exercise 14

Question Number	Answer Grid	Question Number	Answer Grid
14.6		**14.9**	
14.7		**14.10**	
14.8			

Answer Sheet – Exercise 14

Question Number	Answer Grid	Question Number	Answer Grid
14.11		**14.14**	
14.12		**14.15**	
14.13			

Answer Sheet – Exercise 14

Question Number	Answer Grid	Question Number	Answer Grid
14.16		**14.19**	
14.17		**14.20**	
14.18			

EXERCISE 15: Multi Part Numerical Reasoning

Instructions: Work out and mark your answers by filling the appropriate grids on the answer sheet. Use a separate sheet for your workings. *Maximum time allowed: 15 minutes*

QUESTION – ONE

The circumference of a circle is calculated using the formula 2 Π r, where Π is 3.14 and "r" is the radius.

The area of a circle is calculated using the formula Π r²
An inch is the same as 2.54 cm.

Use these facts and formulae to answer the following questions.

15.1 The radius of the wheel of Joel's motorbike is 25 inches. Find the circumference in cm rounded to the nearest unit.

15.2 What is the distance travelled in km, if the Joel's wheel goes around 2500 times? Round to the nearest unit.

15.3 The diameter of a circle is 50cm. Find the area rounded to the nearest unit.

15.4

Find the area of the circle below.

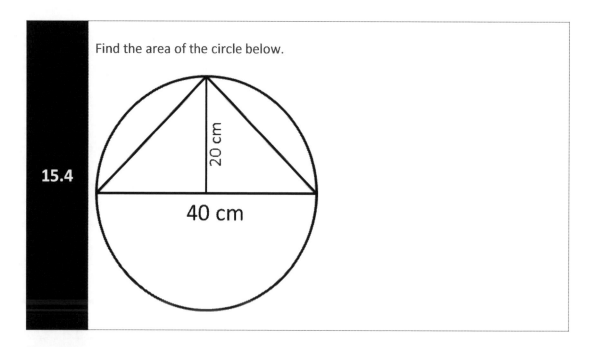

20 cm

40 cm

15.5

A straight line of length 125.6 cm has been curved to form a perfect circle. What will be the radius of the circle?

Using the facts below, find the answers to the first three questions.

The cost of three bananas and an apple is 90p

The cost of three apples and a banana is £1.10

15.6 What is the price of 4 bananas and 4 apples in pence?

15.7 What is the price of one banana in pence?

15.8 What is the price of 10 bananas and 10 apples? State your answer in pound.

15.9 What is the price of one apple in pence?

15.10 What is 60% of 6000?

The symbols ♍, ♎, ♓, 📫 and ♌ each stand for one of the digits 1, 2, 3, 5 or 9 but not necessarily in that order. Use the clues below to work out what number each symbol stands for.

$$♎ \times ♎ = ♓$$

$$📫 \times ♎ = ♎$$

$$📫 + 📫 = ♍$$

$$♍ + ♎ = ♌$$

15.11 What is the value of the symbol ♍?

15.12 What is the value of the symbol 📫?

15.13 What is the value of the symbol ♎?

15.14 What is the value of the symbol ♌?

15.15 What is the value of the equation below?
$$♍ + ♎$$

The questions below are not related to each other.

15.16 Danish was born on 1st May 2004 and Ibrahim was born on 1st March 2011.
By how many months is Danish older than Ibrahim?

15.17 A shopping bag and its contents weigh a total of 3 kg. The bag on its own weights 250 g. The shopping bag has the following contents:

- 2 bags of apples, each weighing 500 grams
- 1 bag of bananas weighing 3/4th of a kilogram
- A bag of oranges

Find the weight of the bag of oranges in grams.

15.18 Three and a half dozen eggs weigh 2520 g. How much would eight dozen eggs weigh in grams?

15.19

Find the value of "x" in the following magic square.

14		7	2
x		12	
	5	9	16
15			3

15.20

A jar with 8 chocolates in it weighed 160 g. The same jar weights 304 g with 20 chocolates in it. How much does the jar weigh on its own?

Answer Sheet – Exercise 15

Question Number	Answer Grid	Question Number	Answer Grid
15.1		**15.4**	
15.2		**15.5**	
15.3			

Answer Sheet – Exercise 15

Question Number	Answer Grid	Question Number	Answer Grid
15.6		**15.9**	
15.7		**15.10**	
15.8			

Answer Sheet – Exercise 15

Question Number	Answer Grid	Question Number	Answer Grid
15.11		**15.14**	
15.12		**15.15**	
15.13			

Answer Sheet – Exercise 15

Question Number	Answer Grid	Question Number	Answer Grid
15.16		**15.19**	
15.17		**15.20**	
15.18			

ANSWER MATCHING
NUMERICAL REASONING

EXERCISE 16: Numerical Reasoning – Answer Matching

Instructions: The table below lists the answers for the questions in this exercise in a random order. Match the correct answers to each question. *Maximum time allowed: 10 minutes*

A	B	C	D	E
12	24 cm^2	700	12	£27.50
F	G	H	I	J
32	525 ml	£8.00	24	32 cm

16.1 What is $^1/_8{}^{th}$ of £64?

16.2 What is the value of "x" in the equation below?

$$x^2 - 44 = 100$$

16.3 Half of 8^2

16.4 What is the total surface area of a cube whose length is 2 cm?

16.5 What is **0.024 x 10^3**?

16.6 What is **35% of 1.5 litre?**

16.7 An adult ticket to a cinema cost £5.50. What's the cost for 5 adults?

16.8 If East is 3, West is 9 and South is 6 then number indicate North?

16.9 The radius of a circle is 16 cm. What is the diameter?

16.10 $\dfrac{350}{0.5} =$ [?]

Answer Sheet: Exercise 16

16.1	A	B	C	D	E	F	G	H	I	J
16.2	A	B	C	D	E	F	G	H	I	J
16.3	A	B	C	D	E	F	G	H	I	J
16.4	A	B	C	D	E	F	G	H	I	J
16.5	A	B	C	D	E	F	G	H	I	J
16.6	A	B	C	D	E	F	G	H	I	J
16.7	A	B	C	D	E	F	G	H	I	J
16.8	A	B	C	D	E	F	G	H	I	J
16.9	A	B	C	D	E	F	G	H	I	J
16.10	A	B	C	D	E	F	G	H	I	J

EXERCISE 17: Numerical Reasoning – Answer Matching

Instructions: The table below lists the answers for the questions in this exercise in a random order. Match the correct answers to each question. *Maximum time allowed: 10 minutes*

A	B	C	D	E
56	330	10	136	13.2
F	G	H	I	J
45,000	61	6.33	12	50

17.1 How many minutes in 5 ½ hours?

17.2 What is 30% of 150,000?

17.3 A car is travelling at 50 miles an hour. How many hours does it take to cover a distance of 600 miles?

17.4 The following sequence follows a pattern of $n + n^2$

2   6   12   20   30   42

What is the next number in sequence?

17.5 What is $10^4 \times 0.00132$?

17.6 What is $6^2 + \sqrt{625}$?

17.7 A square has a perimeter of 200 cm. What is the length of each side in cm?

17.8 What is 64p + £5.69?

17.9 What is the value of "x" if "y" in the equation below is 5?

$$2x^2 + y = 205$$

17.10 A train journey starts at 10:13 AM and ends at 12:29 PM. How many minutes did the journey take?

Answer Sheet: Exercise 17

17.1	A	B	C	D	E	F	G	H	I	J
17.2	A	B	C	D	E	F	G	H	I	J
17.3	A	B	C	D	E	F	G	H	I	J
17.4	A	B	C	D	E	F	G	H	I	J
17.5	A	B	C	D	E	F	G	H	I	J
17.6	A	B	C	D	E	F	G	H	I	J
17.7	A	B	C	D	E	F	G	H	I	J
17.8	A	B	C	D	E	F	G	H	I	J
17.9	A	B	C	D	E	F	G	H	I	J
17.10	A	B	C	D	E	F	G	H	I	J

EXERCISE 18: Numerical Reasoning – Answer Matching

Instructions: The table below lists the answers for the questions in this exercise in a random order. Match the correct answers to each question. *Maximum time allowed: 10 minutes*

A	B	C	D	E
£2.33	£3.43	5°C	180°	1000
F	G	H	I	J
3.0	-7°C	£7.00	5	£5000

18.1	Potatoes are What is the cost of 3 kg of potatoes and 2 kg of carrots?

18.2	Two internal angles of a right angled triangle are 90° and 45°. What is the total of all the three angles?

18.3	Jamie has £10 and spends £7.77 in a shop. How much money is remaining?

18.4	What is 7% of £49?

18.5 The annual take home salary of Rashmi is £60,000.
What is her monthly take home?

18.6 In a long jump competition, John jumps 4m, 3.5m and 4.5m in the first three attempts. The fourth attempt is ruled out due to a foot fault. What is the mean of the four attempts?

18.7 The temperature at midnight was -3°C. It dropped by another 4°C by 3:00 AM. What is the temperature at 3:00 AM?

18.8 The highest temperature today is expected to be 7°C higher than the lowest, which is -2°C. What is the expected high temperature?

18.9 How many sides are there in a Pentagon?

18.10 How many 2p coins are required to make £20?

Answer Sheet: Exercise 18

18.1	A	B	C	D	E	F	G	H	I	J
18.2	A	B	C	D	E	F	G	H	I	J
18.3	A	B	C	D	E	F	G	H	I	J
18.4	A	B	C	D	E	F	G	H	I	J
18.5	A	B	C	D	E	F	G	H	I	J
18.6	A	B	C	D	E	F	G	H	I	J
18.7	A	B	C	D	E	F	G	H	I	J
18.8	A	B	C	D	E	F	G	H	I	J
18.9	A	B	C	D	E	F	G	H	I	J
18.10	A	B	C	D	E	F	G	H	I	J

EXERCISE 19: Numerical Reasoning – Answer Matching

Instructions: The table below lists the answers for the questions in this exercise in a random order. Match the correct answers to each question. *Maximum time allowed: 10 minutes*

A	B	C	D	E
15	39.41	25	91	3600
F	G	H	I	J
5	£300	10 cm	20	8

19.1 What is the height of a triangle whose area is 25 cm^2 and the base is 5 cm?

19.2 A new office tower being built is planned to have 30 floors with each floor catering for 120 desks. What is the total number of desks in the building?

19.3 How many seven-seater vehicles are required for a group of 30 people to travel?

19.4 What is half of 78.82?

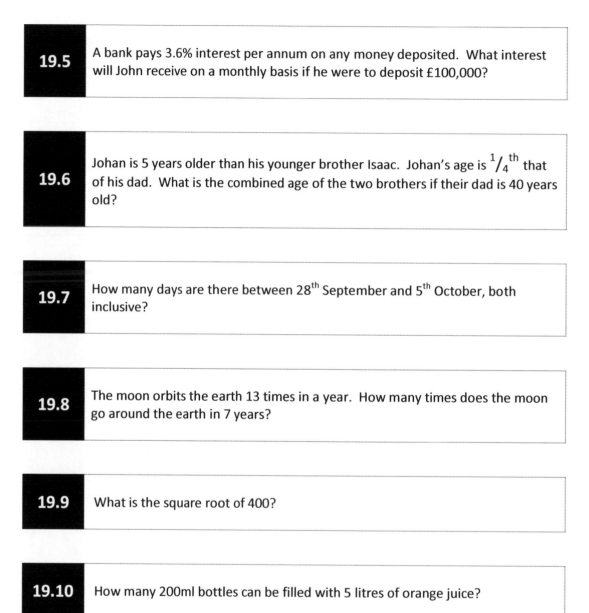

19.5 A bank pays 3.6% interest per annum on any money deposited. What interest will John receive on a monthly basis if he were to deposit £100,000?

19.6 Johan is 5 years older than his younger brother Isaac. Johan's age is $\frac{1}{4}$th that of his dad. What is the combined age of the two brothers if their dad is 40 years old?

19.7 How many days are there between 28th September and 5th October, both inclusive?

19.8 The moon orbits the earth 13 times in a year. How many times does the moon go around the earth in 7 years?

19.9 What is the square root of 400?

19.10 How many 200ml bottles can be filled with 5 litres of orange juice?

Answer Sheet: Exercise 19

19.1	A	B	C	D	E	F	G	H	I	J
19.2	A	B	C	D	E	F	G	H	I	J
19.3	A	B	C	D	E	F	G	H	I	J
19.4	A	B	C	D	E	F	G	H	I	J
19.5	A	B	C	D	E	F	G	H	I	J
19.6	A	B	C	D	E	F	G	H	I	J
19.7	A	B	C	D	E	F	G	H	I	J
19.8	A	B	C	D	E	F	G	H	I	J
19.9	A	B	C	D	E	F	G	H	I	J
19.10	A	B	C	D	E	F	G	H	I	J

EXERCISE 20: Numerical Reasoning – Answer Matching

Instructions: The table below lists the answers for the questions in this exercise in a random order. Match the correct answers to each question. *Maximum time allowed: 10 minutes*

A	B	C	D	E
6	750 m^2	-18	10.5 cm	625
F	G	H	I	J
39483	22.5	125	£400	25

20.1	The thickness of a book is 2 cm. How many copies of the book can be stacked on a shelf of length 2.5m?

20.2	What is the missing number? $45 \times 3 = \boxed{} \times 6$

20.3	What is the surface area of a room, 10 m long and 7.5 m wide?

20.4	Johan gives 10% of his salary to a charity every month. If his annual salary is £48,000, how much money does he give monthly to the charity?

20.5 It takes 30 minutes to paint a wall whose surface area is 2 m^2. How many hours does it take to paint 24m^2?

20.6 What is the diameter of a circle whose radius is 5.25 cm?

20.7 What is the product of 321 and 123?

20.8 What is **3 x 24 / (6 - 10) ?**

20.9 How many times does 0.8 cm go into 5m?

20.10 What is the square root of 625?

Answer Sheet: Exercise 20

20.1	A	B	C	D	E	F	G	H	I	J
20.2	A	B	C	D	E	F	G	H	I	J
20.3	A	B	C	D	E	F	G	H	I	J
20.4	A	B	C	D	E	F	G	H	I	J
20.5	A	B	C	D	E	F	G	H	I	J
20.6	A	B	C	D	E	F	G	H	I	J
20.7	A	B	C	D	E	F	G	H	I	J
20.8	A	B	C	D	E	F	G	H	I	J
20.9	A	B	C	D	E	F	G	H	I	J
20.10	A	B	C	D	E	F	G	H	I	J

ANSWERS

ANSWERS:

Exercise 1		Exercise 2		Exercise 3		Exercise 4	
Q. No.	Answer	Q.No.	Answer	Q. No.	Answer	Q. No.	Answer
1	D	1	F	1	C	1	F
2	C	2	D	2	A	2	C
3	D	3	C	3	B	3	B
4	A	4	C	4	A	4	D
5	D	5	C	5	A	5	F
6	C	6	A	6	C	6	D
7	A	7	C	7	B	7	A
8	A	8	C	8	B	8	B
9	A	9	B	9	C	9	C
10	B	10	B	10	E	10	D
11	C	11	C	11	B	11	E
12	C	12	D	12	C	12	A
13	D	13	C	13	A	13	A
14	D	14	D	14	C	14	B
15	C	15	A	15	A	15	E

Exercise 5		Exercise 6		Exercise 7		Exercise 8	
Q. No.	Answer	Q.No.	Answer	Q. No.	Answer	Q. No.	Answer
1	C	1	B	1	D	1	B
2	B	2	B	2	E	2	D
3	E	3	B	3	A	3	B
4	D	4	B	4	B	4	B
5	B	5	A	5	E	5	C
6	D	6	E	6	F	6	C
7	E	7	B	7	A	7	D
8	D	8	C	8	B	8	A
9	E	9	D	9	B	9	D
10	C	10	C	10	C	10	C
11	A	11	E	11	F	11	D
12	D	12	C	12	A	12	B
13	B	13	A	13	B	13	E
14	C	14	B	14	E	14	D
15	F	15	A	15	E	15	B

	Exercise 9		Exercise 10		Exercise 11		
Q. No.	Answer	Q. No.	Answer	Q. No.	Answer	Q. No.	Answer
1	A	1	B	1	£26,000	16	180
2	D	2	B	2	£5,330	17	60
3	A	3	F	3	£7,200	18	60
4	C	4	D	4	20%	19	60
5	B	5	A	5	£31,400	20	123
6	B	6	B	6	300		
7	C	7	F	7	110		
8	A	8	D	8	140		
9	D	9	B	9	17%		
10	D	10	A	10	180		
11	D	11	B	11	59		
12	B	12	C	12	59		
13	D	13	B	13	63		
14	A	14	B	14	94%		
15	A	15	B	15	43		

Mastering 11+ / MATHS / Practice Book ONE

Exercise 12			Exercise 13			Exercise 14			Exercise 15	
Q. No.	Answer		Q. No.	Answer		Q. No.	Answer		Q. No.	Answer
1	9		1	25		1	22		1	399
2	41		2	90		2	40		2	10
3	90		3	25		3	£5		3	1963
4	45		4	90		4	2600		4	125.6
5	360		5	625		5	6		5	20
6	£80,000		6	2		6	£75,000		6	200
7	£4,000		7	8		7	20%		7	20
8	40		8	24		8	£6,250		8	£5
9	£66,000		9	56		9	£7,500		9	30
10	£6,120		10	24		10	£56,250		10	3600
11	770		11	24		11	£9		11	2
12	215		12	28		12	£90		12	1
13	232		13	-4		13	£35		13	3
14	30%		14	11		14	5		14	5
15	86%		15	-8		15	80		15	5
16	800		16	£90,000		16	1500		16	82
17	16		17	£60,000		17	£3750		17	1000
18	320		18	£1,500		18	160		18	5760
19	434		19	£263		19	49		19	1
20	£9		20	£1,000		20	£245		20	64

Exercise 16			Exercise 17			Exercise 18			Exercise 19	
Q. No.	Answer		Q.No.	Answer		Q.No.	Answer		Q. No.	Answer
1	H		1	B		1	H		1	H
2	D		2	F		2	D		2	E
3	F		3	I		3	A		3	F
4	I		4	A		4	B		4	B
5	B		5	E		5	J		5	G
6	G		6	G		6	F		6	A
7	E		7	J		7	G		7	J
8	A		8	H		8	C		8	D
9	J		9	C		9	I		9	I
10	C		10	D		10	E		10	C

Exercise 20			
Q. No	Answer	Q. No.	Answer
1	H	6	D
2	G	7	F
3	B	8	C
4	I	9	E
5	A	10	J

WHAT NEXT?

Once you are comfortable with the exercises in this book, it is advised that you proceed to practice books 2 and 3 of this series, which have exercises of progressively increasing levels of difficulty.

The three books will cover the breadth of the KS2 syllabus for Maths and hence are recommended for both Grammar and Independent School entrance exams.

Other books in the Mastering 11+ series:

- ➢ English & Verbal Reasoning – Practice Book 1
- ➢ English & Verbal Reasoning – Practice Book 2
- ➢ English & Verbal Reasoning – Practice Book 3

- ➢ Cloze Tests – Practice Book 1
- ➢ Cloze Tests – Practice Book 2
- ➢ Cloze Tests – Practice Book 3

- ➢ Maths – Practice Book 2
- ➢ Maths – Practice Book 3

- ➢ Comprehension – Multiple Choice Exercise Book 1
- ➢ Comprehension – Multiple Choice Exercise Book 2
- ➢ Comprehension – Multiple Choice Exercise Book 3

- ➢ CEM Practice Papers – Pack 1
- ➢ CEM Practice Papers – Pack 2
- ➢ CEM Practice Papers – Pack 3
- ➢ CEM Practice Papers – Pack 4

All queries to **enquiry@mastering11plus.com**